# AVALANCHES
# AND LANDSLIDES

Designed and produced by
Aladdin Books Ltd
28 Percy Street
London W1T 2BZ

Revised and updated edition published in 2002
First published in Great Britain in 1992 by
Franklin Watts
96 Leonard Street
London EC2A 4XD

ISBN: 0 7496 4406 0

Design: David West Children's Book Design
Designer: Stephen Woosnam-Savage
Editors: Fiona Robertson
          Libby Volke
Picture research: Emma Krikler
                  Brian Hunter Smart
Illustrator: Mike Saunders

A catalogue record for this book is available
from the British Library.

Printed in UAE

# *Natural Disasters*

# AVALANCHES AND LANDSLIDES

## JANE WALKER

FRANKLIN WATTS
LONDON • SYDNEY

# CONTENTS

Introduction
5
What is an avalanche?
7
An avalanche begins
8
The white death
10
The Alps
12
What are landslides?
14
Landslide causes
16
Volcanic landslides
18
Disaster strikes
20
Vaiont, Italy, 1963
22
Looking back
24
Are we causing more?
26
What can we do?
29
Fact file
30
Glossary
31
Index
32

500679023

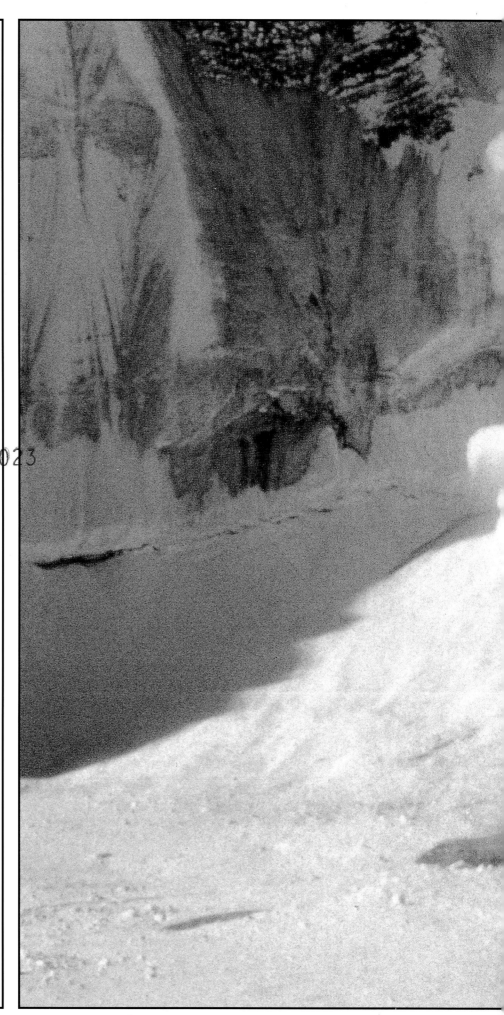

# INTRODUCTION

A huge mass of snow, rock or mud crashes down from a mountain or hillside at high speed. It can engulf and destroy entire towns and villages in just a few minutes. With a thunderous roar, avalanches and landslides strike quickly and often without warning, and their impact can be catastrophic.

Hundreds of avalanches occur each day on remote snowy mountain tops throughout the world. However, their effect is only really felt in the populated mountain regions. Here, the whirling mass of ice and snow will crush to death or bury alive any person or animal unlucky enough to be caught in its path.

The introduction of improved warning systems can reduce the effects of avalanches and landslides. But we must also try to control the human activities that can trigger these natural disasters.

Although the term 'avalanche' can mean the fall of any material – snow, soil, rocks, ice or volcanic lava – it generally refers to a falling mass of snow. Dry-snow avalanches can travel at speeds of up to 360 km/h. The melting snow of a wet-snow avalanche moves more slowly, at about 35 km/h.

## Wet-snow avalanche

Wet-snow avalanches often begin in rocky areas. Here, the snow melts around rocks that have warmed up in the sunshine. The fresh, wet snow can set hard, like concrete, when the avalanche finally comes to a halt.

Melting snow exposes rocky surface

New snow layer

## Dry-snow avalanche

The force of a dry-snow avalanche can equal that of a hurricane. As the powdery snow hurtles downhill, a destructive blast of air may be sent out ahead of the avalanche.

Direction of powder

Air blast

New snow layer

Direction of slab

**Slab avalanche**
As a slab avalanche hurls itself
down a slope, the front of the
slab starts to break up. Slab
avalanches usually occur on
slopes that are protected from
the wind, where the snow
collects in deep piles.

Snow rolls into balls

# WHAT IS AN AVALANCHE?

An avalanche is a huge mass of ice and
snow that breaks away from the side of a
mountain and surges downwards at great
speed. The greatest avalanches probably
occur on the high peaks of the Himalayas.
However, those which cause the highest
death toll fall in the populated valleys
of the Alps.

Scientists have grouped avalanches into
three main kinds: wet-snow avalanche,
dry-snow avalanche and slab avalanche.
Wet-snow avalanches usually occur in the
spring, when the loose, melting snow
forms into large boulders of snow as
it rolls downhill.

More deadly are the dry-snow
avalanches, which either slide along
close to the ground, or lift off the ground
completely and swirl through the air,
often hundreds of metres high. In a
slab avalanche, a huge chunk of solid,
sticky snow breaks away from a slope.
It slides across a layer of loose snow
crystals lying beneath the surface.

# AN AVALANCHE BEGINS

Most avalanches begin either during or soon after a snowstorm. As each new layer of snow settles on the ground, it binds itself to the existing layers that are anchored to the mountainside. The additional weight of a heavy snowfall can prevent the snow from gripping onto the layers of snow underneath, and trigger an avalanche. Avalanches also occur during the spring, when melting snow seeps through the surface. This creates a slippery layer on which the snow can slide.

The steepness of the slope can affect the speed of the avalanche.

Steep, rocky slopes help to anchor the snow. The smooth surface of gentler grass-covered slopes allows the avalanche to reach great speeds.

**Heavy snowfalls**
The weight from deep piles of falling or drifting snow can produce an avalanche.

**Explosions**
An avalanche can be set in motion by an explosion at a mine or quarry, by the noise of a low-flying plane or even by a clap of thunder.

**Avalanches are not only caused by natural factors, such as the amount and quality of snow, or the sudden arrival of strong winds. Once conditions are suitable, an avalanche can be started by the slightest movement, such as a falling icicle or twig, or by a small animal crossing a slope.**

**Cornices**
Hard winds drive the snow and pack it together to form a cornice, which hangs over the very top of the slope. When a cornice falls, it can trigger an avalanche.

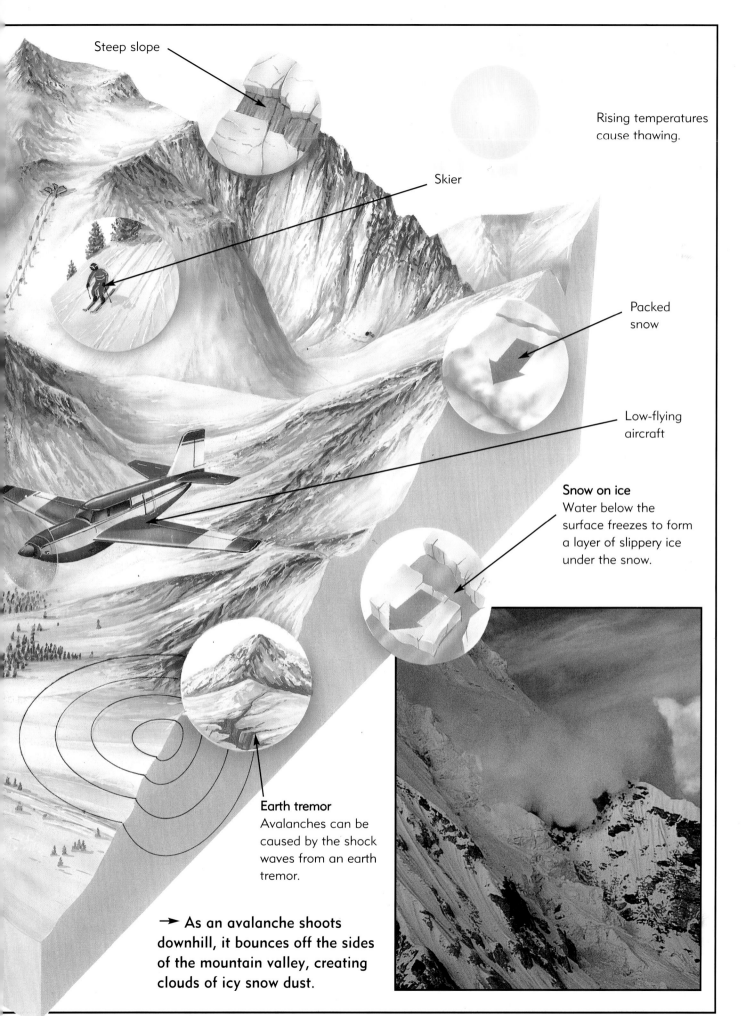

Steep slope

Skier

Rising temperatures cause thawing.

Packed snow

Low-flying aircraft

**Snow on ice**
Water below the surface freezes to form a layer of slippery ice under the snow.

**Earth tremor**
Avalanches can be caused by the shock waves from an earth tremor.

➤ As an avalanche shoots downhill, it bounces off the sides of the mountain valley, creating clouds of icy snow dust.

# THE WHITE DEATH

When an avalanche strikes, more than one million cubic metres of snow and ice blast down a mountainside. As the ice mass falls, it collects large amounts of debris such as rocks and tree stumps on its way. In a populated area, whole villages are crushed, hundreds of people and animals are buried alive, power and water supplies are cut off and roads and railway lines disappear — all in a matter of minutes.

The world's greatest single avalanche disaster occurred in 1962 in Peru. More than 3,500 people died and eight villages and towns were destroyed in just seven minutes. When the avalanche finally came to rest, after a journey of 15 kilometres, the pile of snow and ice was over 18 metres deep.

Trapped under the hard, packed snow, an avalanche victim can barely move and is unlikely to survive more than a couple of hours. Victims die from the cold, a lack of oxygen, or from the injuries that occurred when they were first struck. Only about five per cent of all avalanche victims are rescued alive.

**6.18 pm**
The deadly torrent of ice and rocks misses Yungay, but crushes the village of Ranrahirca, killing 2,700 people.

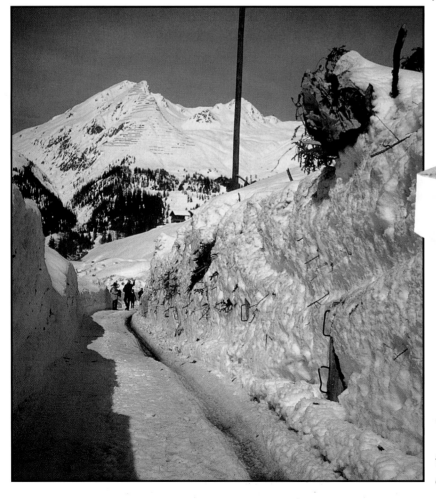

← In this mountain village, a path has been cut through the wall of snow and debris caused by a recent avalanche.

Melting snow plunges down from the north-facing summit of the 6,768-metre high Nevado de Huascarán mountain.

**6.13 pm**
The falling snow loosens millions of tonnes of ice from Glacier 511.

**6.15 pm**
The avalanche flattens Yanamachico and nearby villages, leaving 800 dead and only eight survivors.

The 1962 avalanche disaster in Peru occurred when a huge chunk of ice and snow broke off the country's tallest mountain, Nevado de Huascarán. The avalanche roared downhill at speeds of up to 96 km/h, demolishing nearby villages with a deluge of mud and ice over 24 m thick.

# THE ALPS

Each year, tens of thousands of avalanches occur in the European Alps, the world's most densely populated mountain region. The inhabitants of the numerous Alpine towns and villages live under a constant threat of avalanches, which is made worse by certain natural conditions that prevail there. The warm, dry Föhn wind which blows through the deep Alpine valleys raises temperatures suddenly, causing the snow to melt rapidly. The smooth surface of the Alpine meadows lubricates the path of the avalanches, which travel at terrifying speeds. In 1898, an avalanche at Glarnisch in Switzerland raced downhill at around 400 km/h.

During the 1950-51 'Winter of Terror' in Switzerland, over 1,100 avalanches killed a total of 98 people.

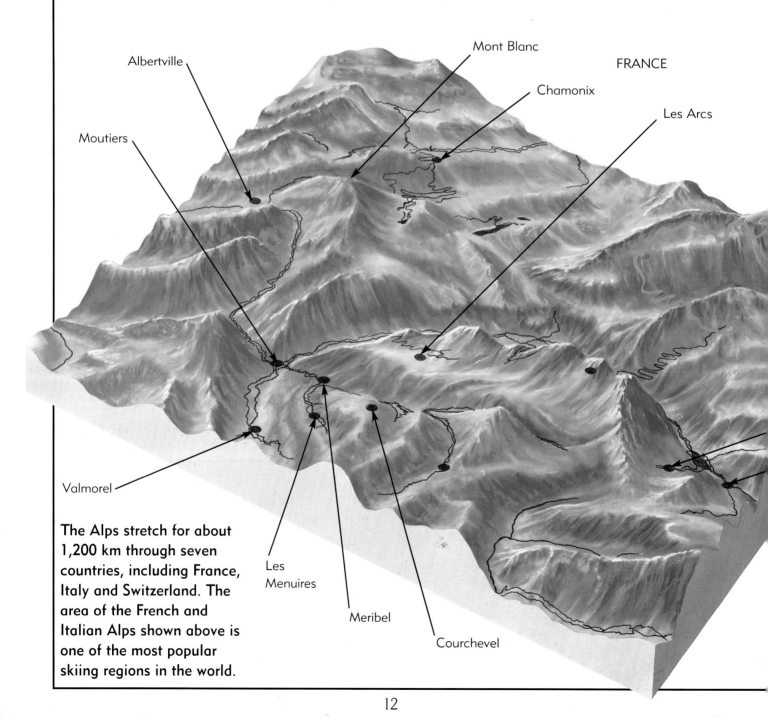

Albertville
Mont Blanc
FRANCE
Chamonix
Les Arcs
Moutiers
Valmorel
Les Menuires
Meribel
Courchevel

The Alps stretch for about 1,200 km through seven countries, including France, Italy and Switzerland. The area of the French and Italian Alps shown above is one of the most popular skiing regions in the world.

↑ Avalanche warning signs, such as the one shown above, inform skiers if there is a danger of avalanches. The ski trails, called pistes, that are at risk are then closed. Experienced skiers who choose to ski 'off piste' on unmarked trails run the risk of setting off an avalanche. The weight of a single skier on a slope can open up a crack in the hard crust on the snow's surface. This releases a huge slab which breaks away, often at a point above the skier, offering little chance of escape.

ITALY

Tignes

Val d'Isère

→ More and more people visit the Alps each year, increasing the likelihood of avalanche fatalities. In 1999, the two ski resort villages of Valzur and Galtuer, in Switzerland, were swamped by deadly avalanches. 38 people died, and 20,000 holiday-makers were stranded for over a week.

# WHAT ARE LANDSLIDES?

A landslide is the name given to the downhill movement of large amounts of soil, mud, rocks and other kinds of debris. There are three different kinds of landslide: falls, slides and flows.

In a rockfall, huge boulders crash down a steep slope, often breaking up into smaller pieces on reaching the ground. When rockslides occur, the chunks of rocky material slide over the ground as quickly as flowing water. Debris slides, which consist of thin layers of loose soil and smaller rocks, travel in the same way.

Mudflows and earthflows are wet landslides. A mass of mud and water flows down from the upper slopes, picking up debris lying in its path.

Curved sliding surface

## Slump

A slump occurs when heavy rock and soil on a steep slope or cliff collapses in a series of curving movements. The blocks of slumped material tilt back towards the slope.

Blocks tilt backwards

## Rockslide

Rockslides involve the downward movement of rock debris or large blocks of rock. They often occur after heavy rains and in areas where large numbers of trees have been removed from the slopes.

Large blocks become detached from slope

**Mudflow**
Mudflows are the most liquid kind of landslide. They occur in mountainous regions, desert areas and on steep-sided volcanoes. They occur after heavy rains and can travel at speeds of up to 90 km/h.

Mudflow streams out of bowl-shaped area

Large tongue of mud, water and fine debris

Rocks break off along lines of weakness

Steep bare slope

Loose debris on slope

**Rockfall**
A rockfall occurs when rocks descend at high speed, often falling freely through the air, down a steep slope or a cliff face. The rocks may have been loosened by frost or ice. On reaching the lower slopes, a rockfall may break up into a moving torrent of debris.

## San Salvador mudslide

In January 2001, an earth-quake hit Central America. It caused a landslide that buried a suburb of El Salvador's capital city San Salvador. A 500-metre landslide swamped the suburb of Las Colinas. 1,200 people died or are missing, and hundreds of houses, cars and streets were lost beneath the mud.

# LANDSLIDE CAUSES

A landslide begins when the loose material on the surface of a slope becomes unstable. Water is one of the key factors in triggering landslides. After heavy rainfall or melting snow, the surface becomes saturated. Water seeps through the top layer, making the layers underneath very slippery. The water increases the weight of the surface material — soil, small rocks and sediment — and weakens the forces that bind it.

As the soil loses its grip on the slope, it can no longer resist the downward pull of gravity. On steep slopes, the water-filled mass forms into a slide. On gentler slopes, it becomes a flow of earth and debris.

Erosion, where the lower part of a slope is cut or worn away, is another major cause of landslides. Erosion can result from natural factors, such as the action of the sea and rivers, or from artificial ones, such as mining and excavating. A landslide itself can be a major force of erosion, especially on steep slopes where the tree cover has been removed.

Landslides are also associated with earthquakes and volcanic activity. After the devastating earthquake of 1964 in Alaska, in the United States, much of the destruction resulted from landslides.

↓ **In the Trisuli Valley in Nepal (below), the soil has been eroded from the steep mountainsides. This erosion is partly due to tree-felling operations in this area.**

The river that flowed through the valley wore away the lower slopes, leaving the upper ones without support.

Water-soaked layers of shale help to lubricate the landslide.

→ A landslide has torn away the ground from beneath these houses in Antofagasta, Chile. They are left clinging to the scarp, or cliff, formed by the slide.

Sandstone

Direction of landslide

Slope of valley before landslide

Debris from landslide

Position of valley floor before landslide

# VOLCANIC LANDSLIDES

On 18 May 1980, Mount St Helens volcano in the Cascade Range of the United States erupted violently, following an earthquake beneath the volcano. As steam, hot gases and other volcanic material shot out from the volcano, the shockwaves from the earthquake triggered off two enormous landslides on the northern side of Mount St Helens.

The falling rocks mixed with melting snow and ice, and the ash and dust from the volcano, to form a kind of mudflow called lahar. The name 'lahar' originally referred to a volcanic mudflow in Indonesia, but it is now used to describe volcanic mudflows anywhere in the world.

Advance warnings had led to the evacuation of the area around Mount St Helens. But five years later, the inhabitants of Armero and other towns and villages lying in the shadow of the Nevado del Ruiz volcano in Colombia were not so fortunate. In November 1985, a violent eruption of the volcano left around 25,000 people dead and 25,000 more without shelter.

The mudflow from the volcano eventually covered an area of more than 40 square kilometres. It destroyed the valuable farmlands of this rice- and coffee-growing area.

↓ A lahar, or volcanic mudflow, can travel over great distances at high speed. Heat from inside the volcano helps to melt the snow and ice on its slopes. The water mixes with loose material on the volcano's slopes to form a massive tongue of mud.

↑ Several weeks before the earthquake, Mount St Helens showed signs of increasing volcanic activity. A large bulge which had appeared on the volcano's northern side was torn apart by the power of the eruption. Clouds of ash and dust poured into the atmosphere.

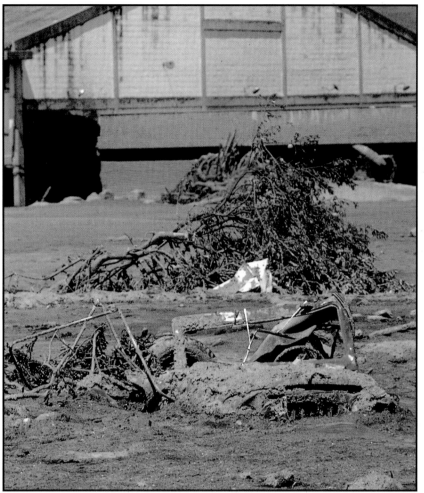

↑ Following the 1985 eruption of Nevado del Ruiz, torrents of mud raced through the streets of Armero at around 160 km/h. Cars were swept away, trees were uprooted and many of the town's inhabitants were engulfed by the tide.

# DISASTER STRIKES

When a landslide begins its descent, huge quantities of soil, mud, rocks and smaller debris hurtle down a hillside. Whole villages are buried, and their inhabitants and livestock are swept along on a tide of mud and debris. Trees are uprooted, houses are flattened or swamped and communications are cut off.

↑ In 1999, torrential rains triggered massive landslides on Mount Avila, Venezuela. Tonnes of debris thundered down into the capital Caracas, burying thousands of people.

The mountainous regions of Nepal in Asia are prone to torrential rains. Here the water-saturated soil has been eroded, exposing layers of rock beneath. This erosion is worse in areas where large numbers of trees have been cut down. The hillsides are stripped of protection against the falling rain, and of tree roots to anchor the soil in place. Landslides glide over the smooth rock at speeds of up to 20 metres per second.

Following an earthquake in the Andes Mountains in Peru in 1970, a devastating mudflow swept through the Peruvian town of Yungay. Almost 18,000 people were buried alive as a giant wave of mud and debris, about 80 metres high, swamped the town.

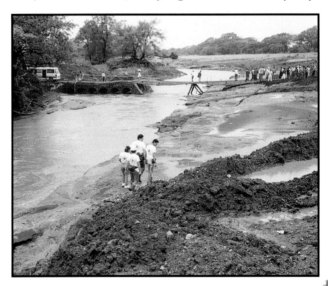

→↑ Rivers of mud swamped southern Italy in 1998, following heavy rains. Worst hit was Sarno, where 80 people died.

→ When the water-filled Casita volcano collapsed in Nicaragua in 1998, a wave of mud killed 2,000 people.

Torrential
rain

San
Gabriel
Mountains

Direction of
mudflow

← In 1969, in the area
around Los Angeles in
the US, almost 112 cm
of rain fell in just 42
days. These rains, some of
the heaviest ever recorded
in the area, washed down
mud and gravel from the
San Gabriel Mountains. As a
swirling torrent of mud, debris
and water swept over the suburb
of Glendora, over 100 people
were drowned and 15,000 people
had to evacuate their homes. The
landslide also blocked the main
highway into Los Angeles.

Glendora

# VAIONT, ITALY, 1963

During the night of 9 October 1963, millions of tonnes of rock slid down from Mount Toc in Italy into the Vaiont Reservoir below.

Three years previously, one of the world's highest arch dams was built below the mountain. Engineers working on the dam had noted that when the water level in the reservoir rose, a rocky mass on the mountainside moved downwards. As the water level increased, so did the rate at which the mass moved. Heavy rains swelled the reservoir,

and the engineers tried to halt the rate of rock movement by lowering the water level. But the rockslide could not be stopped.

At 10:41 pm a torrent of rocks, mud and water raced across the wide river valley and up the opposite slope. As it fell back into the reservoir, a 100-metre high wave was catapulted over the 262-metre high wall of the Vaiont Dam into the valley below.

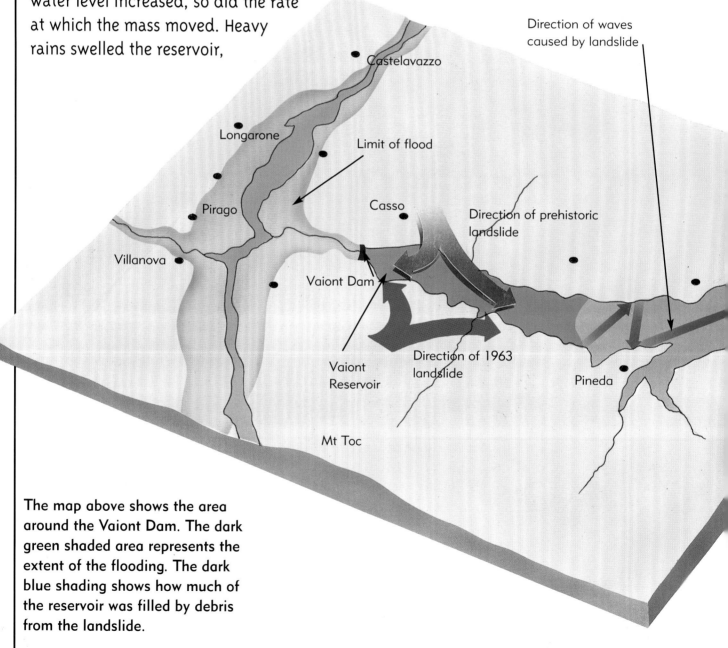

Castelavazzo

Longarone

Limit of flood

Pirago

Casso

Direction of waves caused by landslide

Direction of prehistoric landslide

Villanova

Vaiont Dam

Direction of 1963 landslide

Vaiont Reservoir

Pineda

Mt Toc

**The map above shows the area around the Vaiont Dam. The dark green shaded area represents the extent of the flooding. The dark blue shading shows how much of the reservoir was filled by debris from the landslide.**

Top surface of landslide

The diagram below shows the valley of the Vaiont Reservoir before and after the landslide. In 1960, a landslide fell into the southern end of the reservoir. There is also evidence of a prehistoric landslide near Casso (see map, left).

Limestone

Clay

Beginning of landslide

Direction of landslide

Original shape

← Within two minutes of the rockslide, a flood of water was racing through the narrow valley towards the villages of Pirago and Villanova. It then devastated Longarone (left), where most of the inhabitants died. A total of 2,600 people perished that night.

23

# LOOKING BACK

Evidence of avalanches and landslides dating back to prehistoric times can still be seen today. On the floor of the Saidmarreh Valley, in western Iran, lie the remains of a landslide that scientists believe occurred over 2,000 years ago. Limestone debris covers an area of around 200 square kilometres, and is around 300 metres deep in places.

In 218 BC, Hannibal and his army suffered great losses from avalanches as they crossed the Alps on their way from Carthage to attack Rome. It was October, and fresh snow had fallen on a crust of old snow. Thousands of troops and animals were carried away by the avalanches triggered off as they crossed the unstable crust.

In the Middle Ages, pilgrims from northern Europe regularly crossed the Alps on their way to worship in Rome. Many believed that avalanches were acts of God, while others thought they showed the destructive power of the Devil.

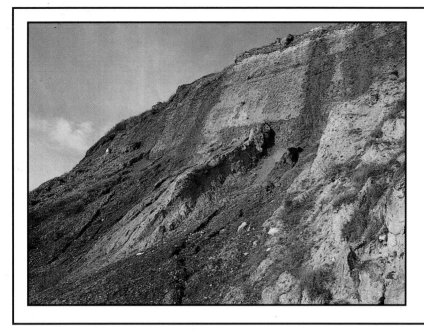

## Landslide!

On the morning of Christmas Day, 1839, a gaping hole opened up in the top of the cliffs at Bindon in Devon, England (left). A vast chunk of rock sank down, leaving an enormous gap about 1,200 metres long and 92 metres wide. The steep slope, or scarp, down which the landslide moved is still visible today, as are the huge lumps of displaced rock.

→ During the winters of the First World War (1914-18), an estimated 60,000 soldiers fighting in the Alps were killed by avalanches. Some of these avalanches were not entirely due to natural causes. After a heavy snowfall, Austrian and Italian troops often fired into the slopes above the enemy below. This action released huge dry-snow avalanches. It was the first time in military history that avalanches had been used as a weapon.

← Hannibal's treacherous journey across the Alps involved about 38,000 foot soldiers, 8,000 horses and over 30 elephants. According to the Roman historian Livius, more than 18,000 men and 2,000 horses were lost during the crossing, mainly because of avalanches. The reduced numbers of Hannibal's troops contributed to their eventual defeat by the Roman army.

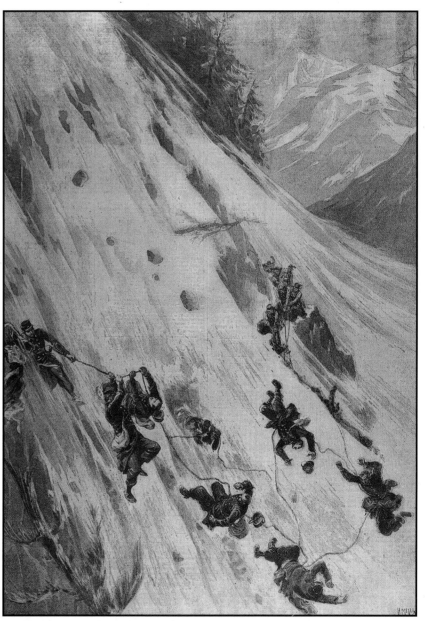

# ARE WE CAUSING MORE?

We cannot prevent natural disasters such as avalanches and landslides. Yet we can stop or reduce the human activities that make it more likely that one of these powerful forces will strike. When large numbers of trees are cut down to provide fuel and timber, as well as land for farming, nothing remains to trap the rainwater and hold the surface of the slopes in place. This deforestation also removes the natural barrier to avalanches and landslides that trees and other vegetation provide.

Mountainous areas in other parts of the world are also being cleared and developed as tourist resorts. Millions of people then visit the mountains to sightsee and explore. These activities can wear away the soil and vegetation, making the area more prone to avalanches and landslides.

**Road construction**
Excavation work for new road-building steepens slopes and removes support for the upper layers.

**Quarrying and mining**
As large amounts of natural resources, such as stone and coal, are removed from the ground, surrounding land becomes unstable and may collapse.

**Deforestation**
The roots of trees and ground foliage hold water like a sponge, releasing it slowly into the surrounding soil in a controlled flow. When a hillside is stripped of its cover, the exposed soil erodes very quickly.

The threat of landslides and avalanches is increased by excavation and construction work, and by the removal of natural resources from the ground. The shape of a slope may be steepened or altered to such an extent that it becomes unstable. Heavy rains and snow add further weight to the slope, weakening the soil's grip on the hillside.

**Reservoirs**
The construction of reservoirs in narrow mountain valleys that are prone to landslides increases the risk of a flood disaster.

↑ Trees help to prevent the erosion of soil by heavy rains. Deforestation has caused this landslide in Malaysia. The slide has completely destroyed a main road.

**Global warming**
As the atmosphere traps increasing amounts of the greenhouse gases that warm the Earth, changes in weather patterns may occur. Very heavy rains could increase the number of mudflows and landslides.

## Forest fires

A lack of hillside vegetation is blamed for the mudflows that are common in parts of southern California in the USA. Forest fires destroy the trees and shrubs on the steep slopes around many residential areas. After heavy rains, the topsoil becomes saturated with water and starts to slip downhill.

← Avalanche fences have been erected on these slopes in the Valais region of Switzerland. These help to keep the snow in place so that no avalanche can form.

Forestation

Snow fences

Roads protected by concrete tunnels

As well as erecting physical barriers against avalanches, concrete tunnels can be built to protect roads and railway lines. Improved warning and evacuation systems also help to reduce the loss of life from these disasters.

Trenches

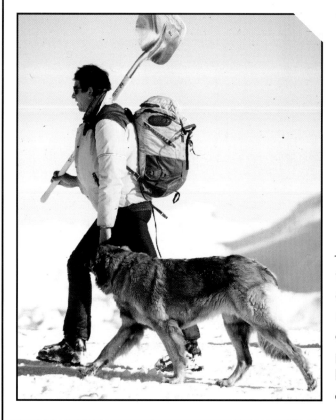

← Following an avalanche, trained dogs use their sense of smell to detect bodies under the snow. Long poles, called sounding rods, are pushed into the snow in the hope of touching an avalanche victim.

Trees
planted

Volcano
early
warning
system

↑ Trees have been
planted on the slopes
above this mountain
village to protect it
from avalanches.

Landslide prevention involves
the use of trenches, pumps and
wells to control water levels.
New developments should be
located in areas where the land
is stable, or where the slopes
can be supported.

Controlled
quarrying

Underground drainage

Landslide barriers

# WHAT CAN WE DO?

Many different measures can be taken
to reduce the devastating impact of an
avalanche or landslide. In Switzerland,
controls include setting off explosives
to release avalanches artificially, and
planting triangular patches of trees
above mountain villages. Anti-avalanche
constructions, such as trees, fences
or barriers, are used to hold the snow
in place, slow down, deflect or break
up an avalanche.

Water is one of the prime causes of
landslides. Measures can be taken to
control and monitor the amount and
pressure of water in unstable slopes.
Underground drainage is also installed
in some places to reduce the water
content in landslide-prone hillsides.
The lower end, or toe, of a slide can be
shored up with concrete or soil.

In the United States, scientists are
using computers to try and forecast
avalanches. The study of volcanoes can
also help to predict volcanic eruptions
more accurately, thus reducing the
death toll from volcanic mudflows.

# FACT FILE

## The world's largest landslide

The largest landslide in the world in the past two million years is believed to have occurred in the United States. At the foot of Mount Shasta, an active volcano in the state of California, lies an area of debris covering some 450 square kilometres. The debris is believed to date from a rockfall that happened between 300,000 and 360,000 years ago.

## RECENT AVALANCHES & LANDSLIDES

### 1990

**Peru** — Following heavy rains, a mudslide buried the village of San Miguel de Río Mayo in the jungle region 800 km north of the capital, Lima. A total of 200 people were reported missing.

### 1999

**Canada** — In the first minutes of the New Year, an avalanche crashed into a school hall in the remote village of Kangiqsualujjuaq, Quebec. Over 400 partygoers were inside. Nine people died, buried under 3 m of snow.

**Austria** — In February, Austria experienced its first avalanche for 50 years. Despite massive rescue efforts, 38 people died in Valzur and Galtuer.

**Venezuela** — Torrential rains triggered horrific landslides and floods across Venezuela's north Caribbean coast and its capital, Caracas. The nationwide death toll may have exceeded 30,000.

### 2000

**Norway** — In January, an avalanche swept a bus and a truck off the road near Kjosen, and plunged them into the sea. At least five people died.

**China** — 213 people died in central China in landslides following July flooding. Thousands of homes were also destroyed and many roads were submerged.

**India** — Also in July, a mudslide in a shanty town in Mumbai buried 45 people. Rescuers scrabbled through the debris with their bare hands to free survivors.

### 2001

**El Salvador** — 1,200 died or vanished in January, when a whole neighbourhood disappeared beneath a landslide triggered by an earthquake.

**Indonesia** — Nearly 100 people died in West Java in February, as landslides entombed them in the makeshift mines they were working in.

**Austria** — A deadly avalanche struck the Tyrol region in February, exactly two years after the Galtuer tragedy. A family of four were crushed in their car.

**Ecuador** — In June, a sodden hillside near Quito collapsed, burying around 40 people taking shelter from the storm.

## Avalanche control in Canada

In Canada's Selkirk Mountains, one of the world's largest anti-avalanche programmes is operated along Rogers Pass, with the help of the Royal Canadian Horse Artillery. In a densely populated mountain area, the soldiers use gunfire to release avalanches and protect a 40-km stretch of major road from 160 separate avalanche paths.

## Mudflow technology

Japan has become a world leader in the study and prevention of mudflows. Since ten per cent of the world's most active volcanoes are found in Japan, many of the country's population centres are threatened by mudflows. On the side of Mount Usu on the island of Hokkaido, steel and concrete structures have been erected to trap falling rocks and slow down the flow of mud and other debris. Elsewhere, television monitors and measuring equipment help to detect volcanic activity and the onset of a mudflow or landslide.

## Volcanic mudflows

On the slopes of Mount Rainier, a volcano in the United States, more than 55 mudflows have been recorded in the past 10,000 years.

## Mining waste disaster

Waste material from mining can be a serious landslide hazard. For years, waste from the coal mines around Aberfan in South Wales was piled dangerously high. When it became saturated with water following heavy rains in October 1966, part of the tip collapsed. The waste material slid downhill, killing over 144 people, including 116 children in the local school.

# GLOSSARY

**avalanche** — the downward movement of a mass of ice and snow.

**creep** — the slow, almost unnoticeable, movement of loose soil down the surface of a slope.

**debris** — loose solid material such as soil, mud, small pieces of rock and stones.

**deforestation** — the large-scale removal of trees and other vegetation.

**dry-snow avalanche** — a huge blast of white powdery snow that travels at great speed along or above the ground.

**earthquake** — a violent shaking movement of the outer layer, or crust, that surrounds the Earth.

**earth tremor** — the violent trembling of the ground that occurs during and after an earthquake.

**erosion** — the removal or gradual destruction of a surface by the action of water, ice or wind. Human activities, such as tree-felling, can also lead to erosion.

**eruption** — the violent release of steam, rocks, dust and ash through an opening in the Earth's surface.

**global warming** — an increase in the Earth's temperature due to a build-up of greenhouse gases, such as carbon dioxide, in the atmosphere.

**greenhouse gas** — gases in the atmosphere that trap heat from the Sun and keep the Earth warm. Greenhouse gases include carbon dioxide and water vapour.

**groundwater** — water that is found in rocks and soil beneath the surface.

**lahar** — a mixture of rocks, melting snow and ice, and the ash and dust from an erupting volcano.

**landslide** — the downhill movement of large amounts of rock and soil.

**lubricate** — to make a surface slippery.

**mudflow** — the rapid downhill movement of soil mixed with water.

**rockfall** — the downward movement, usually through the air, of rocks and other material from a cliff or steep slope.

**rockslide** — the slipping movement of rock along one or more surfaces.

**saturated** — filled with the maximum amount of liquid or solid substance.

**scarp** — the steep slope left behind when a landslide falls away.

**shale** — a kind of soft rock made from tiny pieces of mud and clay.

**shockwave** — a burst of energy that is released from the centre of an earthquake.

**slab avalanche** — the downward movement of a large chunk of solid snow.

**slump** — the movement down a slope of a block of rock and soil. The slumped block tilts back towards the slope.

**unstable** — describes a slope that is likely to start moving downwards.

**volcano** — an opening in the Earth's surface through which hot rocks and gases can escape.

**wet-snow avalanche** — the downhill sliding movement of large chunks of wet snow.

# INDEX

Alps 7, 12-13, 24, 25
anti-avalanche constructions 29
avalanche speeds 6, 8, 11, 12, 31

cornices 8
creep 31

death tolls 7, 10, 11, 12, 13, 15, 18, 20, 21, 23, 25, 30
debris 10, 14, 15, 16, 17, 20, 21, 22, 24, 30, 31
deforestation 14, 16, 20, 26, 27, 31
disasters 10-11, 12, 15, 18, 19, 20-3, 30
dry-snow avalanches 6, 7, 25, 31

earth tremors 9, 31
earthflows 14
earthquakes 15, 16, 18, 19, 20, 30, 31
erosion 16, 20, 26, 27, 31
eruptions 18, 19, 29, 31
evacuation 18, 21, 28
excavation and mining 8, 16, 26, 27, 30
explosions 8
explosives 29

flooding 22, 23, 27, 30

Föhn wind 12
forecasting 29
forest fires 27

global warming 27, 31
greehouse gases 27, 31
groundwater 31
gunfire 25, 30

historical evidence 24-5

ice 5, 6, 7, 9, 10, 11, 15, 31

lahar 18, 31
landslide control 29, 31
landslide speeds 15, 19, 20
lubrication 12, 16, 31

melting snow 6, 7, 8, 11, 12, 16, 18, 31
mountains rescue teams 28, 30
mudflows 14, 15, 18, 20, 21, 27, 29, 30, 31

pistes 13
prehistoric landslides 22, 23, 24
preventative measures 28-9, 30

quarrying 26, 29
rainfall 14, 15, 16, 20, 21, 22, 27, 30

reservoirs 22, 23, 27
road construction 26, 27
rockfalls 14, 15, 30
rockslides 14, 22, 23, 31

sandstone 17
saturation 16, 20, 27, 30, 31
scarp 17, 25, 31
shale 16, 31
shockwaves 9, 18, 31
skiers 9, 12, 13
slab avalanches 7, 13, 31
slumps 14, 31
snow fences 28
snowfall 8, 11, 24, 25, 27

tourist activity 13, 26
tree-planting 29

unstable condition 16, 29, 31

volcanic landslides 15, 16, 18-19, 20, 29, 30

warning systems 5, 13, 18, 28, 29, 30
water 9, 16, 21, 22, 23, 29
wet-snow avalanches 6, 7, 31

## Photocredits

**Abbreviations: l-left, r-right, b-bottom, t-top, c-centre, m-middle**
Front cover l — Digital Stock. Front cover r, back cover — Corbis. 1 — Robert Hardingphoto Library. 4-5, 9 — Science Photo Library. 10bl, 19tl, 27t — Frank Lane Picture Agency. 13t — Lowell Georgia/CORBIS. 13b — Adam Woolfit/CORBIS. 15br — Lopez/Frank Spooner Pictures. 16, 28t — J Allan Cash Photo Library. 17, 19tr — Frank Spooner Pictures. 18-19 — Natural History Photographic Agency. 20tr, 20mr — Rex Features Ltd. 20br — Anticoli/Frank Spooner Pictures. 23 — Tophem Picture Source. 24, 25br — Mary Evans Picture Library. 25tl — Roger Vlitos. 27b, 28b — Spectrum Colour Library. 29t — Geoscience Features.